Soil 'Neath My Feet

SOIL 'NEATH MY FEET

By Robert W. Terrell

VANTAGE PRESS
New York / Washington / Atlanta
Los Angeles / Chicago

FIRST EDITION

All rights reserved, including the right of
reproduction in whole or in part in any form.

Copyright © 1979 by Robert W. Terrell

Published by Vantage Press, Inc.
516 West 34th Street, New York, New York 10001

Manufactured in the United States of America
Standard Book Number 533-03950-9

Library of Congress Catalog Card No.: 78-63046

I hereby dedicate this book to my beloved father
Professor Edson W. Terrell

(April 3, 1896–April 17, 1955)

CONTENTS

Introduction ix
Foreword xv
Prologue xvii

1. Soils and How to Use Them Best 1
    Clay Soils 2
    Loam Soils 4
    Gravel Soils 5
    Sandy Soils 6
    Tillage Tools 7

2. Cover Crops and Pasture Plants 12
    Conservation Practices 12
    Meadow Mixtures 14
    Pasture Mixtures 15
    Ladino Clover 15

3. Pasture Care 18
    Preparation of Soil 18
    Top-dressing and Why 19
    Seed Mixture to Use 20
    Rotation and Why 20
                                            23

4. Meadow Care 23

5. Insects: Their Control and Use 29

6. Benefits of a Farm Pond 33
7. Cattle Feed—Raising and Storage 38
8. Cattle Health 46
9. Calf Husbandry 52
10. Some Problems of the Poorly Educated Farmer 57
11. What to Look for When Purchasing a Farm 62
    Epilogue 68

INTRODUCTION

Throughout the history of man, the farmer has been a highly respected individual; after all, he is the hard-working person who keeps odd hours trying to feed the world and who, in many respects, is responsible for our health and well-being. Unfortunately, the great "luxury" of technology and the use of commercial fertilizers are the primary causes of nutritionally deficient farm products, and the only way to restore and improve the health and environment is through organic farming.

Organic farming, which uses natural materials, has many advantages over the inorganic methods that dominate the country today. Without a doubt, the most important advantage is one that affects the entire populace—health. Long ago, fruits and vegetables, as well as corn, wheat, oats, rye, and other grains, were grown relatively free of disease and pest problems. At that time, however, farmers practiced organic farming, the natural way.

It has been, since the tremendous boom in commercial and mass-producing agribusiness, that the quality of our feed has deteriorated. Even in the early 1800s, a farmer noted that "the lack of persistent, steady effort which soon became an American characteristic, led at once to a superficial and exhausting mode of cultivation which has resulted in reducing thousands of acres of once fertile soil to a barren wilderness." To "heal" this barren soil, many farmers use chemical fertilizers, which in turn poison the produce. Today, the world is burdened with such diseases as cancer, high cholesterol, diabetes, anemia, and, even in this advanced country, incredible deficiencies. People are told that they need not take vitamin supplements or medicines if they eat the right foods; this should be true, but what if the foods themselves are lacking essential nutrients?

Our food is being robbed of its natural protein properties and then fortified with inorganic vitamins and filled with lethal additives and preservatives. Health foods are becoming increasingly popular, but still farmers insist on farming inorganically; they are following practices set for them by their fathers and grandfathers and assuming it to be the right way. Any organic farmer will testify that it is not.

Although not a miracle cure for all ills, organic farming is one of the best (for that matter, one of the only) ways to help save ourselves and the environment. We have all read about obscure villages hidden in foreign countries that possess good mental and physical health. It is reported that these surviving primitive cultures are centuries behind our time; the people eat natural foods and live in stress-free worlds, and, ironically, they seem to be centuries ahead of us! They have what we strive for—health and happiness. Western health standards are incredibly low due to our poor diet; rather than spend the money and time visiting doctors to cure our ills, we should view organic farming as a sort of preventive medicine.

The health of farm animals deserves equal consideration, for these creatures benefit as much from organic farming as people. Often, animals tend to become one of a large group of numbers that often results in a strict culling program due to widespread ill health. With a switch to organic methods, you need never be concerned with such such cattle diseases as mastitis, acetonemia, milk fever, and other animal sicknesses. Wild animals have been equipped with the ability to heal themselves, and a return to nature would spare the farmer these problems. This is a tremendous advantage in itself.

The first consideration in starting or running a farm is finance. It is necessary to be able to manage a business enterprise, for farming can be an expensive proposition if not handled properly. Remember that you will be selling your goods wholesale and buying supplies retail, and a poor manager will lose money rapidly. This business does take money, and a young man who would like to farm exclusively just doesn't stand a chance. To start, you must have a great deal of capital, and general farming demands more than twice the capital layout for machinery than specialized farming does.

For grain and grass farming, special equipment is necessary; to raise and harvest a crop of corn, you would need a corn planter, cultivator, corn harvester and wagons, silo filler and unloader, and a conveyor to bunkers for feeding. Should you also grow wheat, oats, soybeans, hay, etc., you would need still more machinery. In the eastern United States, ample moisture makes grass farming the most reasonable business enterprise. Hay production requires the least amount of capital for equipment when compared to any other specialized farming. If cared for properly, a well-prepared seedbed and good seed mixture will annually produce two heavy cuttings of excellent hay for over thirty years! You can't top that.

It is best, then, to become a specialized farmer; be either a grass or grain farmer but not both. Eastern grain farmers compete with the specialized grain farmers of the Midwest, often fighting a losing battle. Three out of five years, Eastern corn crops never mature, and heavy late-summer storms cause large losses in oat and wheat yields.

Farming organically is more economical and practical. These farmers are not subject to the outrageous prices and inevitable shortages of commercial fertilizer and are employing the best possible method for soil improvement. They are helping nature to restore its own balance with longer-lasting organic materials. Artificial ingredients do not add any beneficial substances to the soil; in fact, they are leaching compounds and are extremely harmful. Products on the grocery shelves have almost all been poisoned at least once by chemicals and preservatives that have been incorporated into these marketable and edible foods during their processing. A decrease in commercial fertilizers will not only eliminate most of these poisons but will save the air and water from pollution attributed to fertilizer factories. The land used for farming will be utilized more efficiently, enabling other areas to be used however they are needed. It is a vicious cycle, and we must act NOW if it is ever to stop.

On my own farm, I have proved beyond the shadow of a doubt that benefits resulting from organic farming are tremendous. Statistics are not necessary; my crops have been healthier, larger, tastier, and more abundant than those of my neighbors, who farm inorganically.

It is hard to understand why the switch from natural farming to inorganic practices occurred so rapidly, yet a return to earlier methods is so slow in taking place. Soon, hopefully, farmers will show large commercial fertilizer and pesticide companies that cooperation with nature produces better results; and what better proof is there than something raised by the two most powerful elements on earth—man and Mother Nature. When these two elements, so often in conflict, work together, the product is bound to be good.

There is a terrific feeling of satisfaction gained from growing something naturally; operating a sort of partnership with nature provides an inner strength and perfection. Inorganic farmers are competing with the environment rather than working with it. By farming organically, you will have plenty of good, wholesome food to eat, lots of good water to drink, plenty of wind, rain, and sun in your face. You have given yourself health and happiness, and that is what really counts.

FOREWORD

In January 1973, I was asked to attend the first "International Convention on Organic Farming and Gardening," being held in San Francisco, California. Its organizer and coordinator, Mrs. Bargyla Rateaver, asked me to write a book on organic farming, which she intended to use as a textbook in her classroom teaching at California State University in Sacramento. This task I have accomplished; but in a broader sense, the material enclosed can and should be read by every person interested in the health of all soil-abiding life and the preservation of our natural resources.

It is my understanding that this is, if not the only one, at most, one of the few books covering such a broad area of organic farming practices.

It is written in minute detail so that the average layman can understand the various methods and procedures used.

Any person interested in using that which was created for our stewardship here on this earth will truly benefit from a lifetime of strict organic agricultural practices that I have indulged in.

It is solely up to you as an individual to carry these practices to their fruition. If they are not strictly adhered to, then you will have been a failure.

PROLOGUE

As I sit down and relax, I ponder about the past events here on my small estate, and I thrill at my accomplishments. All happenings, I can truthfully say, have been and are a work of love. They thrill me as I meditate on them.

It is extremely difficult for me to understand how I could have been farming organically for over thirty-three years while my neighbors and all of my farm acquaintances always farmed inorganically.

My father was a teacher of vocational agriculture husbandry for over thirty years, so I have been reared with a farming background.

Because my father was a teacher of all phases of farming, we were continually experimenting, so we could not help but realize which type of farming we professed to be the best. In my estimation, there is only one way to farm—God's way, the organic way. There is no lukewarm road to follow. You are either hot or cold for organic farming.

Remember that healthy soils produce healthy crops, healthy crops produce healthy animals, better beef and dairy products result in a healthy populace. That statement is so very simple and so very true. Through both greed and competition, our farming methods have so deteriorated that today a very large percentage of our youth are not physically fit for our armed services.

As stewards of the soil, men must practice God's way of cropping the land, *which is the organic way.* After all, everything here on this earth was placed here by God for man's use. We are not here to destroy, as we have been doing, but to utilize and manage what God has provided. Let us establish a program of soil management and then share with our fellow men the fruits of our labor. We are here but a short time. We came into this world with nothing; we leave with nothing.

If we stray from the principles of the true organic way of life, we will assuredly have to suffer the inorganic way of living that man is now following.

It is better to choose the organic way of life because God made it the better choice.

The following is an excerpt from "I Am a Farmer."

I AM A FARMER

I am the provider for all mankind. Upon me every human being constantly depends. A world itself is built upon my toil, my products, my honesty.

Because of my industry, America, my country, leads the world: her prosperity is maintained by me; her great commerce is the work of my good hands; her "balance of trade" springs from furrows of my farm. My reaper brings food for today; my plows hold promises for tomorrow.

In war I am absolute; in peace I am

indispensible—my country's surest defense and constant reliance. I am the very soil of America, the hope of the race, the balance wheel of civilization. When I prosper, men are happy; when I fail, all the world suffers.

I live with nature, walk in green fields under the golden sunlight, out in the great alone, where brain and brawn and toil supply mankind's primary needs; and I try to do my humble part to carry out the great plan of God.

Even the birds are my companions; they greet me with a symphony at the day's dawn and chum with me until the evening prayer is said. If it were not for me, the treasures of the earth would remain securely locked; the granaries would be useless frames; man himself would be doomed speedily to extinction or decay. (Author unknown)

Soil 'Neath My Feet

Chapter One

SOILS AND HOW TO USE THEM BEST

All life, as we know it, is dependent, either directly or indirectly, on the soil. Animal life and mankind are both tied closely to the soil and the vegetation that it produces. This relatively thin layer of material, which makes up only a very small percentage of the earth's crust, is the key to existence on our planet.

Soil comes in varying compositions and consistencies, but all soils can be said to have one basic thing in common: all are a combination of dead and living material, or rock and mineral deposits plus organic material. The rock and mineral makeup of a soil depends on the way in which erosion over centuries has shaped the particles and combined elements to create the mineral; the organic material present in a soil depends on the sort of vegetation that has died and become a part of the soil itself. Soils can range from coarse gravels to fine silts and from extremely acidic soils to those with more basic compositions. Whatever type of soil exists in an area, it can be altered through careful, well-planned management and can thus become more useful and productive.

Since not all soils are alike, each different type of soil requires specialized farming methods in order to get the best results from that particular soil. This seems obvious to the point of being almost trite, but it is alarming how many farmers actually do not understand the soil they are trying to farm and have no idea of the tillage methods that are best suited to their needs. One does not need to be a scientist to understand soil and how to work it; all it takes is a little knowledge and some common sense.

Essentially, there are four different types of soil structures. These are clay, loam, gravel, and sand. They differ primarily in the size of particles in the soil and the distance between these particles. A clay soil, for example, consists of very fine silt particles that are very compact in structure; sandy soils, on the other hand, have larger particles that are grouped in a loose, coarse structure. They are all different in nature, and all require their own different methods of management. The soil contains all of the necessary basics of life. All elements and minerals necessary for our use are present. Organic materials are the only required replacements.

Clay Soils

Clay soils are very heavy and tightly compacted. Because of the closeness and density of the particles, this type of soil readily retains water and humus-building materials. It will not leach away minerals and organic material as easily as other soil types. Because of its compact structure, however, it can easily become a "dead" soil, one that will not produce crops. Unless it is properly managed, and the correct tools are used to work it, it will not be as productive as it can potentially be.

Tools used in the management of a clay soil should be of the chisel type, such as a subsoiler, chisel plow, field cultivator, and the spring-tooth harrow. All of these tools will aerate the soil and thoroughly mix any herbage and manures through the topsoil where crops will be able to make the best use of these natural plant foods. Clay soils are so tightly compacted that unless a subsoiler or chisel plow is used, rainwater will be unable to penetrate the surface and reach the roots of the crops; it will collect in useless (and even harmful) pools on the surface unless the clay is broken up. During hot and dry weather, an improperly cultivated clay soil will become hard and smooth as cement and totally useless to the farmer. A clay soil that is not aerated properly will also tend to be too dense to permit a plant's roots to penetrate it. This soil type should never be worked in any manner during a wet period. More harm than good will result.

Aeration is really the only major problem in farming a clay soil. Since these soils are usually found on level and low-lying areas—and seldom on hills or graded regions—they are not subject to erosion by wind and water the way other soils are. Besides, the compact nature of the clay itself fights erosion and holds the soil in place. If the soil is properly aerated, it is really one of the best types with which to work. It is the ideal soil for maintaining permanent meadow or pasture lands but can be adapted to all sorts of crops and grains simply through the use of the correct tools and tillage procedures.

Loam Soils

A farmer who is fortunate enough to have loam soil to work with has a distinct advantage because loam soils are the best all-around soil structures to manage. Loam has a firm body because its particles are not as compact as the clay types not as loose and porous as the gravels and sands. Loam is actually a mixture of clay, sand, and organic matter. It is usually very dark and rich-looking. When at its best, it will crumble in your hands.

Since loam consists of larger particles than clay, and these particles are more loosely structured, it does not require the extensive aeration techniques needed to farm clay soils. The coarseness of the soil permits adequate penetration of the heat, light, and moisture, all of which are essential to a plant's development. However, the coarseness of loam also allows for speedy evaporation of ground water, and these soils tend to dry out more quickly than clay soils. In regions of abundant rainfall, this is not a problem, but farmers in drier climates must know how to prepare the soil to allow for maximum use of available water.

A subsoiler should be used to break up the hardpan, the tightly compressed subsoil. This permits moisture, humus, and heat from the sun to penetrate the topsoil and reach the root level of the plants in the subsoil.

The ideal manner in which to use a subsoiler is to drag it in all directions across a field, up and down and back and forth. If, however, the farmer does not have time to do this, he can simply drag his subsoiler back and forth across the field, spacing the cuts approximately three feet apart. The subsoiler is the greatest drainage tool ever developed; in heavy soils, such as clay and loam, it permits ground-level moisture to penetrate through the subsoil; this means that heavy rainfalls, which would otherwise be unable to soak in and therefore run off, causing sheet erosion and much loss of topsoil, can penetrate the subsoil and remain where they can benefit the crops and increase the water table. Today's farmer has enough problems facing him without having to worry about erosion, and erosion can actually be avoided through careful soil management. However, subsoiling should only be practiced in moderately dry weather where it will fracture the soil in all directions. In wet weather, it will only make a slice through the soil.

Other tools used in preparing a seed bed in loam soils are as follows: disk plows, moldboard plows, disk harrows, cultimulchers, and cultipackers.

Gravel Soils

Gravel soils are very coarse in structure, containing particles of rock and pebble and often mixed with clay. Because of its coarseness, it is very porous and can lose large amounts of moisture and plant nutrients through evaporation and erosion.

Gravel soils, because of their structure, are very tough on tillage tools, and the coarse particles can easily wear away the cutting face of the tools.

In spite of its shortcomings, gravel soil can easily

become a good productive soil through the right management techniques.

Some advantages offered by gravel include its resistance to weeds and its excellent drainage capacities.

Weeds do not pose as competitive a threat to cultivated plants in a gravel soil; as a rule, only the plants that are intended to grow in gravel really thrive, and the weeds do not reach a level on which they deprive the crops of nutrients or water.

Gravel soils can be cultivated earlier than other soils in the spring because of the fine drainage qualities they possess. Eastern farmers can often till and plant their gravel soils as early as mid-March, while most clay soils cannot be prepared until the first of June. Clay soils hold the cold and dampness of winter much longer than the more porous gravels, which allow the penetration of heat and light and warm up very rapidly in the spring.

Sandy Soils

Sandy soils can vary in structure: there are sandy-loam soils and pure sandy soils. Pure sand, like that found on the beach, is of little use to farmers, but a soil that combines both sand and loam is an excellent soil to manage.

The only drawback to sandy-loam soils is that they require irrigation. These are porous soils, like the gravels, which lose water rapidly through evaporation. The structure, however, is close enough so that it holds humus-building materials and plenty of heat from the sun for quick seed germination and fast, excellent plant growth. If the farmer keeps the water level up through proper irrigation, a sandy loam is as good as any other soil.

Like gravels, sandy-loam soils are relatively weed free, a great asset to the farmer.

Sandy soils are not particularly good for use in the dairy industry, as the herbage eaten by cattle will contain sand particles that wear down the teeth of the animal in the same way they wear off the cutting edge on tillage tools. This shortens the productive life of the livestock. Because cattle are cud-chewing animals, they cannot function properly once their teeth are worn down; an animal with worn-out teeth is unable to masticate its food for proper digestion.

Each of the four main types of soil can be made productive if the farmer is willing to determine which soil he is working with and follow these general guidelines for their management. Crops planted in correctly prepared soil have at least a proper beginning and are potentially of high yield and good quality. A farmer who chooses to remain ignorant about the soil is really doing nothing more than wasting his time and money; he is in the wrong business. His crops will never reach their full potential, and his farm will be doomed to mediocrity or failure.

Tillage Tools: How to Prepare the Soil

Tillage tools are the secret to successful farming. Through good tillage practices, a farmer can alleviate the need for costly and harmful fertilizers. If the seed bed is well prepared at the proper time and season, there is no need for the use of chemicals.

The sad truth is that most farmers do not know what machinery they need and what they don't need. A high percentage of American farmers today have far more machinery than they really need, and the tools and machines they do have are in many cases not suited to the job they face.

The most important single farm implement is the tractor because all other modern machinery is dependent on its performance. The tractor is both the source of power and motivation for the tillage tools used with it.

When buying a tractor, it is important to get the right size for the farm in question. Too many farmers buy oversized tractors, which cost more and eat up far more fuel than a smaller one, even though the smaller one is often better suited to the job. A farmer who buys an oversized tractor is paying for power he'll never even see in use.

For most farms, a medium-sized tractor, one of about 35-40 horsepower, is plenty. Anything bigger is waste. And one tractor per farm is all that's needed unless your farm is abnormally large and requires the simultaneous use of several tractors to keep up with the work. It is simply amazing how many farmers buy two or more tractors rather than one versatile one; many farmers use one tractor for certain jobs, while the other sits idle, using a wasteful and senseless "job rotation" for their machines.

If one tractor is used, as it should be, the drawbar power of that tractor should be just powerful enough to handle the heaviest work on the farm; the smaller jobs, then, will not be as costly, as there will be less wasted energy.

Along with the tractor, the most important pieces of equipment are the subsoiler and the field cultivator. These two are indispensable to the organic farmer.

Use of these two tools in preparing the seed bed forms the foundation of the farm.

The subsoiler is used to break up or fracture the hardpan, which lies several inches beneath the surface of the topsoil. If this harder layer is not penetrated, moisture, heat, and oxygen will not be allowed to reach the root level of the plants, and the roots themselves will have difficulty reaching the necessary nurients. They can starve to death in the midst of plenty.

After the subsoiler has been used, the field cultivator further works the soil by mixing it to a depth of fourteen to twenty inches. During this process, any refuse or cover crop on the soil will be mixed thoroughly with the soil base. A field thus prepared acts as a sponge during wet weather and will retain vast amounts of rainfall; if dry weather comes, this water reservoir is invaluable.

These two tillage tools, the subsoiler and the field cultivator, are the most important pieces of machinery to the farmer. They are needed in all kinds of soil.

In today's farming enterprise, the tillage tools used in preparing a seed bed for the planting of any type of herbage are chiefly manufactured for tractor use. The advance from horse power to tractor power has revolutionized agribusiness in the past forty years.

Machinery today is much larger in size and weight, and with the automatic lifts built into tractors today, transportation and manipulation of this equipment are relatively easy. Imagine a team of horses drawing a disc weighing over 2,000 pounds, a set of eight or ten plows on one beam, or a field cultivator with nine or more chisels with a penetration depth of eighteen inches or more! Farm machinery dealers have racks in their display rooms containing brochures showing and explaining the use of all types of farm equipment used with tractor power.

There are several categories of farm machinery, but the three main categories that farmers are interested in are soil conditioners and planting and harvesting machinery.

This country is well covered with implement dealerships that handle all brands of farm machinery and accessories.

Depending on the size of tractor power, a farmer chooses his farm equipment. The size of your tractor is very important because you can be overpowered, and that makes your tractor expensive to operate when you are using it on smaller jobs. The drawbar power of your tractor should be just stout enough to handle your heaviest farm work; then it would make the smaller jobs cheaper.

Farm-implement dealers usually handle a specific brand of tractors and equipment. The sensible thing to do is to purchase your farm tractor and the necessary equipment that is manufactured for its use. Also, it is extremely important because you can get it serviced much sooner, thereby saving valuable time when time can be of the utmost essence.

Most farmers are not only overpowered with size and numbers of tractors, but they are also overburdened with too much equipment. Most of this equipment just sits around and rusts away. These farmers are equipment poor. Their overhead is much too high, and their investment is staggering. Management is the secret here. One farmer needs only one tractor, one field cultivator, one disc, one harrow, etc. He can only operate one tractor at a time, so what good are two tractors? This will lower his investment and his overhead. There will be fewer repairs, less depreciation, and he should have more capital to work with.

Chapter Two

COVER CROPS AND PASTURE PLANTS

A cover crop is any type or variety of herbage that will produce a sod after planting. Cover crops are vital to organic farming because they protect and rebuild the soil, returning to it many organic nutrients and elements that may have been lost through the growth process.

After a crop has been harvested, the soil is barren and unprotected. As such, it is exposed to the weather and liable to erosion by wind and rain. It is because of the lack of good cover crops that many farmers in grain-producing states have consistently lost so much good topsoil to dust storms. The soil must be protected and nourished during the off season in order to be ready for planting the following year.

Two types of land most vulnerable to wind and water erosion are sloping or hilly lands and lands in elevated locations. In such locales, the farmer has two basic options regarding how he will use his land: contour farming and strip farming.

Contour farming is used when farming land that is hilly or rolling. Basically, it is nothing more than farming in contours *around* the slope of the hill in a screwlike pattern rather than planting in furrows straight up and down the slope. This simple procedure cuts the loss of topsoil by water runoff.

Strip farming is most often used on elevated fields where the land is subjected to frequent winds. The field is planted in strips, alternating between the desired crop (corn, for example) and the cover crop, consisting of grasses and legumes. In this method, the alternate strips of cover plants protect the tilled strips from both wind and rain and help hold the precious topsoil in place.

The width of these strips depends on the steepness of the slope being farmed. If the field has steeply sloping places in it, the strips of cover planting should be wider; this assures that excess rainfall will not simply flow over the protective strips and carry the topsoil away. Wide strips on sharply contoured land are the best way of preserving the soil. In areas where the incline is not so drastic, the strips can be fewer and narrower.

What are the best types of plants to use for cover crops?

Crown vetch is considered the king of cover crops. You have probably seen it planted along the banks of superhighways. It is a vinelike plant that entwines itself among other plants to form a thick entanglement of foliage that holds the soil tightly in place. It grows at an incredible rate, and a few vines will quickly cover a large area. In fact, government authorities—unaware of the aggressive character of crown vetch—are now encountering unexpected difficulties with the vine along our nation's highways, as the vetch continues to grow over everything in sight. They got more than they bargained for.

The major difficulty with crown vetch is finding seed for it. The government has bought up most of the existing seed supply for its highway-conservation usage, and the average farmer finds it nearly impossible to obtain. Such incredible prices as twelve dollars for twenty-four plants have been charged due to the shortage. In spite of all the expense involved, crown vetch is still the best ground cover available.

Vetch is a legume, and the fernlike foliage produces delicate pink and purple flowers that last from June until the first hard autumn frost. It is a hardy plant, resistant to disease, and will thrive in even the poorest of soils. When mature, it forms a dense mat of foliage that chokes out even the most persistent weeds and provides excellent erosion control to steep slopes and banks where nothing else can be grown.

Next to crown vetch as a cover crop I would recommend bird's-foot trefoil. It is an excellent herbage to recover from either cutting for hay or pasturing with livestock. Its leaf area starts close to the ground and therefore permits its fast and luxurious growth. It also spreads quickly and soon becomes a thick and excellent hay crop.

Perennial rye grass is an excellent cover crop to both renovate worn-out soil and to use in a temperate climate where a harsh but open winter takes place.

It is a fast-growing grass and will produce a heavy sod that is a wonderful soil conditioner. Plant it in the month of September and it will make an excellent protective covering before winter sets in. In the spring, you can utilize this crop in your farming program as you see fit.

If seed can be found at reasonable prices, the ideal mixture for seeding permanent meadow would combine eight pounds of Climax timothy seed, four pounds of Penn Gift vetch seed, and eight pounds of Penn Lake orchard grass seed to each acre of ground. These plants are companionable, that is, they will not destroy each other by competing for light, water, and minerals, and produce an excellent hay crop.

If vetch is unavailable, a good alternative cover crop can be formed by mixing five pounds of Viking Bird's-foot trefoil (a three-leafed clover), eight pounds of Climax timothy seed, and eight pounds of Penn Lake orchard grass seed. Although not as good a cover as vetch, the trefoil will protect the soil and yield excellent hay.

Another frequently used clover is Ladino clover. This is a fine plant, but improper management of it can cause the farmer many problems.

In the first place, it is very easy to overdo it when seeding Ladino clover. The seeds are small, and amounts can easily be miscalculated. Ladino should not be used in excess of one-quarter pound per acre because the actual plant when mature is so large and hardy that it often crowds out other pasture plants. Use it sparingly for best results.

Another common complaint against Ladino clover is that it can be a major cause of bloat in cattle. If cattle are to be pastured on lands covered by Ladino, they should be fed dry roughage, such as legume hay, before they are put out to pasture. This reduces the chances of their incurring bloat from the clover roughage they eat in the pasture. The dry roughage will displace a large amount of the succulent clover, thereby preventing the cattle from overeating and becoming bloated.

If you should happen to have a field that has been mistreated in the past and has a considerable amount of hardpan, a good fast-growing plant that will mellow the soil and also add a great deal of humus to the soil in one growing season is a grain called buckwheat. In fact, you can disc the first crop into the soil and grow a second crop the same growing season. This is a wonderful crop for preparing the soil for whatever crop or legume you want to plant later. A cautionary note, however, is apparent. Do not let either crop bloom and produce seed or the crops you plant later will be polluted with buckwheat plants.

There are scores of grass and legume seeds for sale by commercial outlets. You should be careful when purchasing seed for your immediate area as some companies will sell you a "shotgun" mixture that will contain seed grown in a much warmer climate than yours, and, therefore, the young plants will winterkill, and you will have spent your money for naught.

Purchase only seed that is adaptable to your area. The type and condition of soil you are farming has a great bearing on the type of pasture and meadow plants you should be interested in establishing. Because two cuttings for hay are a general practice, you should seed your meadows and pasture land to mixtures that will not depend on reseeding themselves annually.

Keep your seed mixtures simple, but do use a mixture of grasses and legumes. You will have less trouble helpwise and gain more tonnage per acre over a longer period than you would with a seeding of one single grass or one single legume.

Cover crops are too often taken for granted, yet they are one of the most important aspects of successful agribusiness. They protect the topsoil, eliminate weeds, and provide the farmer with his own cheap cattle feed.

A farmer who uses his cover crops wisely will not have to waste his money buying hay and grain supplements for his livestock. He has plenty of both growing right there on his own land.

For young farmers who do not understand the use of mouldboard plow in the soil, I will explain to them that if a fourteen-inch plow is being used, every field plowed with a grade or contour is moving that whole field downhill fourteen inches. If this is kept up year after year, you can imagine where your soil will end up.

Hilly ground should always be kept in sod. Once properly tilled and seeded, all you need to do is either top-dress it with manure semi-annually and make hay here or else keep it in permanent pasture.

If you persist in plowing hilly ground, you are going to end up with a poverty farm in time. Nature will not cooperate with a farmer and a plow on hilly ground.

Chapter Three

PASTURE CARE

A pasture is a very important part of any farmer's enterprise. In spite of this, many farmers are terribly negligent with their pasture, allowing it to grow up in underbrush, thorn apple trees, and wild willows; they are often so bad that a person cannot even walk across the field. This land, which is frequently the poorest section of a farmer's land, should not be forgotten. It should be cared for to the same extent as meadow land and should, by all means, be kept productive; after all, this area is taxed just as heavily as neighboring fields.

It is natural for healthy soil to be slightly acidic. However, if the pasture is composed of the poorest land on a farm, and if it has grown up to underbrush, the soil may become acidic due to negligence, overcropping, overgrazing, or lack of organic materials being returned to the soil. The soil has been literally leached of organic materials. Most farmers would have the area analyzed for acidity and would follow with a heavy application of lime, as prescribed by the test results. However, liming soil in an acidic state would be similar to a person taking bicarbonate of soda for an upset stomach; the lime will sweeten the soil by neutralizing the acid content. This acidic condition is concentrated in the top few inches of soil; therefore, by tilling the soil to a depth of 10–12 inches, the condition can be easily controlled. Also, by top-dressing the field fairly heavily with manure the first growing season, a farmer can control both the acidic and alkaline condition of his pasture area.

Manure mixed with bedding is a farmer's gold mine; through good management, a farmer can realize a permanent pasture with livestock alone. It is surprising how many farmers farm from a sack (commercial fertilizer), while tons of manure stand idle and are leached through exposure to the weather. Often, three or four years' worth of manure are stacked and covered with sod; although there is a substantial loss of nutrients and other essential elements, these piles can be classified as crude compost and may be used as sheet compost and in the same way as raw or green manure. A farmer should never use too much fresh cattle manure on his fields because of possible contamination, resulting in poor quantity returns and off-flavored milk. It is more advantageous to top-dress the field often with a lighter application of green manure.*

You should be extremely careful when top-dressing any field on your farm with green manure. The season of the year, the lay of the land, and the type of weather have a great deal to do with polluting the water table on yours and adjoining land.

If a farmer top-dresses on rolling ground that is icy, and it rains, the manures will drain off the hills into nearby streams. On flat land (level) that is too wet the manures and fertilizers will leach away. Plan your

*green manure—fresh cattle droppings.

sheet composting to fit the lay of the land and the weather conditions present.

Pasture renovation should be handled in the same manner as a grassland or meadow. If it has grown up in underbrush, it first should be cleared with a bulldozer and then disked with a set-angle disk or a field cultivator. After a proper dressing of manure, the field should be harrowed and seeded with a proper pasture seed mixture. If pasture fields are accessible by machinery, the same mixture of grass and legume seed can be applied that is used for meadows. However, it is important to remember that it takes good land management and planned grazing if one is to retain the original seeding, thereby having a good nutritious livestock feed for the entire growing season. It is possible to have such a succulent pasture growth that it can be harvested for an early crop of hay.

If you treat your pastureland as you do your meadowland, you will be able to harvest a crop of hay from it three out of every five years. It is surprising how great the yield of roughage can be from pastureland properly cared for.

One great mistake most farmers make is to turn their cattle out to pasture too early in the spring. The young plants do not have a chance to get the proper growth and therefore have no recuperating powers. The cattle will keep well ahead of the growth, and the pasture will become nothing but an exercising yard for the remainder of the growing season.

If rotation in grazing (sectional grazing) is desired, an electric fence for dividing the field is advisable. This type of fence is less expensive than any other kind and is more easily maintained, especially in areas where frost and heavy snowfalls pose a threat. Brand-new four-strand barbed-wire fences can be ruined in one winter season by heavy snow, for as the snow settles, it stretches the wires and pulls the posts. In the spring, the fence would have to be repaired or replaced.

A Page wire fence, usually constructed as a boundary barrier, is another good type. It has at least one strand of barbed wire to stabilize the fence and prevent livestock from reaching over and breaking it down. However, it is more expensive than the electric kind and also is vulnerable to the winter elements. On some farms or ranches, a board fence is used and is quite attractive. It, too, is susceptible to winter weather and requires a lot of extra upkeep, such as painting or covering with a wood preservative.

Industrial fencing is used the least because of its very high costs and is not affordable by the average farmer; initial cost and installation charges are both quite high. It is easy to see, then, the advantages of electric fencing. Not only is it much cheaper, but also field sections or plots can be controlled by a single strand.

Another aspect of pasture care involves the livestock. They are important because of their organic manure, which enriches and builds the soil's fertility and capability to produce a profitable harvest. By keeping the cattle confined in a barn or enclosure during the night, grazing can be controlled, thus lengthening the grazing season and providing more nutritious roughage for cattle. This also keeps the seeding from dying out and protects the animals during inclement weather. At the same time, the farmer has all the extra manure deposited during the night to place or spread where it is needed most. The livestock can also be fed hay, an excellent and safe supplement to good pasture grazing.

An average-size herd of cattle kept in the barn each night during the pasture growing season will provide enough manure to properly top-dress around thirty extra acres of land.

Inorganic chemicals upset the soil's balance and disturb the mineral and chemical content of the crops; this directly affects the health of animals that feed on those crops. With proper fencing methods and nightly controlled cattle care, the livestock will be both productive and healthy. Through organic soil care, utilizing only manure, fields will be kept in continuous and productive use for many many years.

Chapter Four

MEADOW CARE

A cow's main source of nourishment is in the roughage she consumes, particularly roughage in the form of hay. To a farmer with a herd of dairy cattle, who specializes in grasses and legumes as his main source of roughage for the animals, the care and maintenance of cattle feed from his meadows is very important. Therefore, the better the quality of roughage harvested from the meadow, the more profitable the results will be. A recipe such as silage or soft corn, as either a supplement or major part of cattle feed, requires a substantial investment in machinery and storage facilities. Overhead from these expenses can rapidly ruin the farmer. Furthermore, the hay from the meadow is preferred by cattle over *concentrates*. Concentrates include both grains and legumes, such as soy bean oil meal, cottonseed oil meal, linseed oil meal and alfalfa meal, fed as a livestock supplement; too much concentrate will upset the cow's digestive system, just as candy does to a child. It is most important to realize that the health of farm livestock has deteriorated in the same manner as human health.

A cow has four stomachs, requiring a lot of dry

roughage during the off-growing season to satisfy the animal. If a farmer manages his cattle and meadow well, he will feed hay year-round.

Nature will never permit a field to lay barren; it will produce wild herbage and, later, brush and trees. This is a natural prevention of wind and water erosion. In starting a meadow, it is important, first of all, to realize that through good, basic land management practices, today called "conservation measures," the farmer can even further nature's goals for soil protection. Farming land in vertical strips or lines on slopes will cause tremendous erosion problems, while farming with the contour of the land offers protection from destructive natural elements. Another soil-saving method employs strip farming, which entails planting alternating strips of grasses and grains. Remember these practices when preparing a meadow—or any farm land; they are picturesque, yet they still adhere to good farm management.

Although this may surprise some farmers, the best method for preparing a profitable meadow does not utilize crop rotation. Most of today's farmers do rotate crops, as, perhaps, did their predecessors for years previous before. This farming method is not progress, but it *is* expensive and old-fashioned and has many disadvantages. Most farmers, using a four-year rotation system (i.e., hay, corn, oats, etc.), would drill grass seed with, for example, the oats since oats grow quickly; the grass seed must compete for available moisture and nourishment in the soil, as well as for sunlight. Furthermore, weeds are always a problem when new seeding is done frequently. Each time new ground or an old meadow is broken, a new crop of weed seed is brought to the surface, thereby producing and distributing a competitive crop that would not exist in a field in permanent meadow. Granted, a weed is only a misplaced plant, nature's way of soil and wildlife care; however, a heavy weed crop in the meadow would no longer be a protective force, but a destructive one. What a farmer is striving to obtain, after all, is a good seeding for his meadow; a good stand of new seeding in a four-year crop rotation system can become very expensive.

A permanent hay harvest in spans of twenty to forty years is far less expensive than rotation and, since any field preparation or seeding is costly, requires the utmost care in soil treatment. A farmer should plan ahead and know exactly how he wants to achieve his permanent meadow. The condition of the seed bed is very important for several reasons. The bed should be smooth and granular, yet firm, since this will retain more moisture and will not dry out quickly should there be an unforseen dry spell. In addition, the seeds being drilled are small and should not be covered with more that one inch of soil; seeds with harder shells will lie dormant for several years in the soil if they are planted too deeply.

In midspring, weather permitting, harvest a crop of hay. The field should be top-dressed with manure for the following reason. Plants can and do grow in soil where trace elements are not readily available, but in order for them to mature to their fullest nutritional value for animal feed, they must be grown in completely fertile soil. Natural manures provide a complete and balanced plant food. Therefore, cover the field with a medium manure coverage, anywhere from eight to twelve tons per acre. Harvest a second crop of hay in August. The land should then be plowed, disked, and harrowed to perfection (as described in Chapter Two) and seeded with the preferred seed mixture. No nurse or cover crop is used, so that the new seeding has little competition from a mere scattering of weeds. It will grow extremely well throughout the remainder of the growing season and provide excellent ground cover during the winter. In the spring, when harvest season begins, the new seeding can be mowed as usual with no loss of hay crop to rotation or new seeding, which receives a thorough beating while a nurse crop is being harvested. This way, there have been no interruptions in the farming schedule.

In curing hay, the early part of the harvest season often can be difficult due to inclement weather, especially here in northwestern Pennsylvania. Spring rains, which are frequently heavy and prolonged, can create exasperating problems when meadows are ready for harvesting. Therefore, the first crop of grass (hay) should be mowed when about a foot in height and can be left on the field. This will delay the harvest until late spring when rains have ceased and will take advantage of a more adaptable weather pattern to create a higher-quality roughage due to later seasonal cutting.

Whenever a field of hay is harvested, the blade of the mowing machine should be raised and the shoes set as low as possible, thus raising the cutting blade so stubble remaining after the herbage is cut will be at least six inches high. This stubble, or lower part of the stalk, is not very palatable, yet gives the parent stalk a better chance to recuperate and grow to full maturity in a much shorter time; this also presents a higher protein-rated feed. This second crop of hay harvested from the same field is superior to the supplemental concentrates fed to the livestock.

If the farmer does not concentrate on moving, raking, baling, and hauling his hay as fast as weather permits, he will not be able to harvest it at its highest nutritional value. He then will have to feed more concentrate to his livestock in order to replace the protein loss resulting from overly matured crops. Hay that loses its "youth" is not producing the expected tonnage, or has its seeding crowded out by less advantageous herbage can be renovated so that a farmer can realize a cash crop without using a nurse crop in a new seeding. Mentioned earlier was the use of oats not only as a cash crop but also as a nurse crop for seeding permanent meadow. In midsummer, when the oats are harvested, heavy machinery is used that smashes down the new grass seeding while these plants are young and tender. Then the straw from the oat crop must be raked, baled, and hauled from the field, creating havoc for the future meadow. It is easy to see what a rough time seeding in a meadow has when crop rotation is employed.

Remember the poem "Little Boy Blue . . . The sheep's in the meadow, the cow's in the corn"? It is picturesque but not at all practical. Livestock should never be permitted to graze in a meadow because they crop the herbage too close to the soil, thereby destroying its chances for recuperation. It will, of course, destroy all of the weaker plants and permit foreign or wild herbage to gain a foothold; this will drastically limit the tonnage of herbage harvested per acre. Most farmers using the rotation system allow their livestock to graze in the meadow. However, it is very costly and should not become a standard practice.

Once the hay is harvested, storage becomes a problem. Rather than invest in silos for both corn and hay storage, it is more economical to construct an addition on to the barn for extra storage. Years of study show that the only sound reason for building an upright silo is inclement weather. In places such as the Great Lakes, there is little fear of drought; moisture in abundance is a blessing, as opposed to its scarcity. If the weather is not adaptable to curing roughage in the field, it can be harvested and stored in a trench or pit silo.

In summary, the four basic steps for expertly handling a successful new seeding for permanent meadow are:

1. Be sure the seed bed is properly prepared.
2. Use the best obtainable seed for your region.
3. Use only organic fertilizers (manure).
4. Do not use a nurse crop, it is too competitive.

It is easy to see how growing good, healthy roughage in a permanent meadow has many advantages over the other systems of farming practices.

Chapter Five

INSECTS: THEIR CONTROL AND USE

Insects can destroy a crop as fast an any natural calamity or misuse of the land. They can also be of great use to the farmer who knows how to control and manage them.

First of all, not all insects are harmful, and it is important for the farmer to realize this and be familiar with the species and habits of insects he encounters on his land.

There is an element of balance built into the system of nature by which insect pests are readily controlled by other insects and birds. Man, through his use of poisons and pollutants, has changed that balance.

The availability of poisons for insect control has brainwashed farmers. They now think in terms of these poisons exclusively and find it easy to simply spray their fields with contaminants to get rid of the unwanted bugs. Poisons will, of course, kill off the unwanted insects; what people tend to forget—or prefer to ignore—is that poisons do not stop at one particular species of insect. Poison kills nearly all insects and often birds and other animals as well. A farmer may get rid of aphids or beetles that have infested his crop, but he is also exterminating thousands of natural predators that live by eating the harmful insects.

In short, the farmer who dumps poisons over his crop is doing nothing but harming himself. If he had enough common sense to think beyond that one crop he may save with poison, if he could peer just a short distance into the future and see the overall effects of his inorganic farming methods, he would probably stop the practice immediately.

This is just another illustration of the stupidity of inorganic farming. It is a stupidity that replaces common sense with expediency and encourages farmers to do certain things simply because they are easier and more effective in the short run. Long-term consequences are ignored, but they can no longer be overlooked

Do you realize that already, after a relatively short usage of poisons for insect control, certain insects have developed immunities to our poison? It no longer kills them. So what do we do? The obvious: we simply brew up stronger and stronger poisons. They may be effective for a while, as they kill off harmful and useful insects at the same time, but it can only be a matter of time before the insects become immune to these stronger poisons as well.

Just close your eyes and imagine a future in which vast fields of crops lie rotting and useless, devoured by insects that no poison can kill. It's not as inconceivable as it may sound. In fact, it's starting to happen already. What will we do, then? Perhaps develop a super poison that nothing can withstand and wipe out everything?

There is no question that the continued use of poisons can lead to nothing but bad results. Not only insects and birds are affected, but also the animals who eat the contaminated crops—and this includes human beings.

What is the alternative to toxins? What can we do to rid ourselves of the insect pests that threaten our crops. The answer is simple, so simple that I'm sure most advocates of the modern, "sophisticated" agribusiness will scoff at its simplicity. "Farm organically."

Get rid of your ignorance about the soil, all your fertilizers, poisons, chemicals, and additives and farm in accordance with nature. Rely on common sense and hard work rather than on the propaganda of inorganic sources and the multimillion-dollar industries that stay in business by selling pollutants and useless products.

Insects thrive on diseased plants. Healthy ones are relatively safe from damage by insects. Most of the diseases that afflict our crops stem from inorganic farming methods. It thereby stands to reason that a field of healthy, organically prepared plants stands a better chance of resisting insects than a field of sickly, poisoned ones.

By not using poisons, a farmer protects the birds and insects that can rid his fields of pests. Lacewing, trichogramma, ladybird beetles, praying mantises, these are only a few of the natural predators that prey on harmful insects such as aphids, beetles, and caterpillars.

I know from first-hand experience that natural predators and healthy crops, combined, are enough to ward off insect pests. I know because I have seen it working for over thirty years on my own farm while my neighbors pour thousands of dollars of expensive poisons over their crops and still are plagued with pests. I not only save money but also help preserve wildlife and raise healthier crops. And people still argue in favor of inorganic farming!

Remember that a healthy plant emits an odor that will repel harmful insects. They have their own built-in immunity: so why should any farmer ever want to forsake a "Garden of Eden" for the type of farming methods used today?

If you ever have the time, stop at your garden center store in town and read the labels on containers containing insecticides and herbicidal sprays and powders. Maybe then you will be convinced by the use of such deadly chemicals in agriculture there has to be destructive results.

In the insect world, there are both harmful and beneficial insects. What we call the good insects are the ones that feed on the harmful insects, so why, if there is this balance in nature, should we try to upset it? Every time man interferes with nature, he seems to upset the normal pattern, which generally backfires and causes more trouble. The trend should be to cooperate with nature instead of oppose her! So, by farming organically, we are cooperating with her and will benefit both financially and healthwise. You must see to believe.

Chapter Six

BENEFITS OF A FARM POND

"A pond is one of nature's crucibles. The water in it contains a segment of life's spectrum, both narrow bands of one-celled organisms and wider bands of creatures more complex."

The preceding quote describes a mysterious attraction to the pond, a small body of water that may seem bothersome or insignificant to some but is actually an important asset to the success of a farm.

There are many ecological and economical reasons a farm pond should be included, if at all possible, in plans for organic farming. Almost every farm has acreage with either a slight, natural depression that can be enlarged or a shallow ravine where a dam site could be constructed to form a pond. To enlarge a natural depression involves a greater removal of earth, creating a larger area in which to build a dam; this, in turn, results in more labor and money. However, the benefits are well worth the investment. The ideal foundation for a pond is in a ravine or hollow, which is already wasted land and would be best utilized in this fashion.

Even with today's high machinery and labor costs, a farm pond one-half to several acres in size does not require the preparation one may envision, and in a few short years, it will more than pay for itself. For example, the United States Department of Agriculture estimates that about three-fourths of the water that is used on farms comes from surface sources (streams, lakes, and ponds) and one-fourth from wells and springs. As a result, the value of the farm property will increase.

The primary purpose of your farm pond (i.e., for insurance, recreation, fire protection, or livestock watering) will determine its most feasible location. Stake out an area of not less than one-fourth acre and plan on a minimum depth of eight feet because of evaporation; remember that the deeper the water, the colder the temperature. There are, incidentally, several types of ponds. A pond originating from a natural spring is a very common kind (especially in Pennsylvania) and may actually end up being deeper than eight feet. Two other sources may be a diverted flowing stream and a pond fed by surface runoff only, a type that is very unpredictable in its water level and supply.

After selecting the pond site, remove all trees, brush, and large rocks from the area, being certain to leave some stones for fish protection. The line of the dam should be staked out to indicate the length of the dam and the area of the water basin. It is essential to know the dam's height and width before beginning to build; one cannot construct a farm pond haphazardly and expect it to serve its purpose.

Use a bulldozer to remove topsoil and pile it outside the dam area for later use, not to use as fill. Prepare a spillway, both a permanent one as well as one for emergency overflow to protect the dam from heavy rains and possible flooding. A spring-tooth harrow should be used to finally prepare the ground.

Since this is to be a pond on an organic farm, it is crucial to remember *not* to utilize any artificial fertilizers in or around the area. The fish and other pond life require food and a suitable environment, thus needing some sort of fertilization to start plant life in the pond. With a manure spreader, top dress what will be the pond bottom with manure; this should be done again the second year to assure rich growth.

The dam itself should be wide at the bottom and slope up to the top, and the pond itself has a basin effect.

Now that the pond is pretty well constructed, the topsoil removed earlier should be taken to the top of the dam and around the pond for seeding grass. It is best to plan to build your pond in the spring or summer months so that grass will have the opportunity to become well established before winter. The best type of grass I have used for my farm pond is Reed's canary grass, a tall, thick-growing grass, or any grass that can be grown around water.

Presumably, you will want to stock your pond with fish for both recreational and ecological purposes. Pond owners are most frequently urged to fill ponds with large-mouth bass and pan fish, such as bluegills, sunfish, etc. Following the above instructions carefully, fish should be supplied with the suitable environment for breeding. Also, you will have constructed your *own* pond and not one that had to be built to government specifications and stocked with its fish. It will be your very own, private pond and a valuable contribution to the safety, balance, and beauty of your farm. A pond provides a large volume of water for fire protection, thus permitting lower insurance rates on farm buildings.

Other assets for a farm pond include raising the water table in the immediate area and providing water for irrigation purposes. It has a cooling effect on the surrounding atmosphere due to evaporation of water during hot summer days and nights. Farm ponds also attract fur-bearing animals, thereby bringing your farm in even closer contact with the helpful creatures that contribute so crucially to the chain of farm life. In addition, for people who like to trap or hunt, the pond acts as a sort of bait to lure these animals.

Ponds may be used to water livestock when no other source is available. Here, however, care must be taken to see that organic farming is employed and that absolutely no artificial pesticides or fertilizers are "accidentally" washed into the water. A ravine or depression is a natural place for such poisons to collect. Runoff water could cause contamination if it enters the pond; ponds can, at times, show measurable amounts of pesticide residue from water leaving fields under certain conditions. Domestic livestock represents an important part of the farm. Both humans and animals occupy the paradoxical position of both contributing to and being harmed by pollutants added to water, and water is one of the major transmitters of poisons to humans. This is often transmitted via livestock that, after poisoning their own systems with inorganic substances, poison humans who use livestock products. Therefore, the purity of water consumed by animals has far-reaching implications. Polluted water may make cows ill or even kill them, and milk production is also bound to be affected.

There is no real need for concern if organic farming is practiced. Organic materials will contribute to the life chain of a pond; in turn, creatures attracted to a pond will contribute to the creation of organic substances. Stocking a pond with fish will serve several purposes in areas of food, recreation, and ecology. Recreational reasons are rather obvious, but the ecological ones are not so clear and much more important. Fish often serve as valuable water quality indicators and are especially crucial in ponds that may, somehow, collect pesticides. If fish live in ponds used to water livestock, toxic concentrations of pesticides will diminish greatly.

Ponds also attract frogs and toads that use this area to perpetuate their species. During their stay at the pond, these creatures will move about the farm area, eating both good and bad insects, thereby acting as a *natural* pesticide in man's constant battle with nature.

If not convinced by the significant ecological benefits from a pond, a farmer should be easily swayed by the sheer enjoyment of this small body of water. It provides an excellent source of recreation, furnishing a place for swimming in the summer and skating during the winter. It is a haven for animal, bird, and insect life and can provide a sort of picture window for watching nature's methods at work. A pond is almost a natural sedative: a nice place to sit and relax, to contemplate the next chore to be done on your organic-oriented farm.

Chapter Seven

CATTLE FEED — RAISING AND STORAGE

All members of the livestock family are creatures of habit. They travel the same paths each day and group with certain individuals or herds of their own kind; they even follow a timetable. Usually at the same time each day, these animals graze in certain fields, drink water, and, depending on the amount of herbage available, lie down to chew their cud (digest the morning's intake of food). The entire process is repeated later in the day. Cattle are repetitious in their daily life patterns and become upset when subjected to drastic or immediate change in their routine. They are considered dumb animals, obeying only their instincts. It requires a great deal of time and patience to change any part of this routine and often, with certain animals that will not respond to change, it is impossible.

    A cow has four stomachs and is a cud-chewing animal, like all split-hoofed animals with the exception of the pig. A cow does not have any upper front teeth and, in the process of eating, wraps her tongue around the herbage and literally tears it off, using her molars to grind the grass enough to let her swallow it. Members of the bovine family are able to regurgitate their food in wads and properly masticate it before it traverses through the other three stomachs en route to final digestion. A cow is like a factory in itself. She eats grass and manufactures it into milk and meat; she furnishes materials that man later processes into butter, cheese, dried milk, and many by-products. This animal is truly important to man and especially to organic farmers.

A cow will always satisfy her own bodily needs before producing any extra materials for her owner. Therefore, she should have present at all times the best available roughage that the farm can provide. It is imperative that the farmer prepare his meadow and pasture fields with the utmost care, purchasing the best suitable seed for his area, and wisely choose the most advisable planting season.

Seldom will you find two farmers using the same varieties and seed mixtures on their fields, just as they use different models of machinery, different breeds of cattle, and have various methods of conducting their business. With continual experimentation and the use of strict organic practices, meadow and pasture fields seeded with permanent grass and legumes can still produce an excellent and heavy harvest for twenty to thirty years. For such results, it is necessary to fit the field and select a choice mixture of grass seed (or some variety that is an old standby), top-dress with manure biannually, and harvest the crop at its most productive stage of maturity. Through controlled grazing and good management practices, a farmer can use the same grass and legume mixtures on both pasture and meadow land. One recommended mixture contains four pounds of Canadian or northern-grown bird's-foot trefoil, called Viking. Add to this six pounds of Climax timothy seed and six of Penn Lake non-bunch-type orchard grass seed. Remember that trefoil seed, as well as all other legumes, must be coated with a special inoculant or nitrogen-fixing bacteria. Each time the field is reseeded, an inoculant for the special type of legume seed must be used. In the four-year rotation system, this reinoculation process is usually not necessary.

The following is a recipe for the inoculation of legume seed: First, dampen the seed with water and sprinkle enough cornstarch on to form a coating on the seed. Thoroughly mix the inoculant, which is a black sooty substance purchased with the seed at any seed store or cooperative, with the coated seed until it begins to dry. The inoculated legume can then be mixed with the rest of the seed mixture. Do not proceed with this practice until ready to drill the previously prepared field.

If the water table is excessively high, it would be advisable to use Reed's canary grass for seeding in place of legume seed. This grass is adaptable to both wet and heavy soil, produces a good grade of hay, and is an excellent grass for silage.

In raising good grasses for your livestock, storage is also an important consideration; once the feed has been grown and harvested, why ruin it with poor storage? If both grasses and legumes are raised for silage as a supplement for hay, two types of storage may be used. The upright or standing silo is the oldest method and is still the most popular. There have been as many improvements in upright silos as in types and styles of barns. The very oldest type, the wood-stave silo, cannot compare with the huge and rugged steel or cement silos of today; it could not stand the pressures of being filled with today's heavy stands of legumes and the speed that is used to fill them. They are too dangerous to climb inside for a short time after filling because there are deadly pockets of gasses formed by certain herbage mixtures. Once in a while, they may even explode; however, this does not generally happen on a smaller farm where time is not a crucial matter. Men have been known to climb up and enter the silo to check its capacity and die immediately from the noxious gasses. In addition, a large upright silo with automatic unloader and feeding station will cost as much as some farms; twenty thousand dollars will not cover the initial cost with installation!

The trench silo can be constructed in a pit, slight depression, or right on top of the ground. It is much more reasonable in price, is safer, and holds as much silage as the farmer needs because it is on top of the ground. There are several ways to protect and preserve the silage in the trench silo. In past years, farmers covered the silage with any available material, such as soil, sawdust, hay, etc. Today, however, they use a plastic sheet held in place with old automobile tires; it is economical and makes an air-tight seal, thus assuring minimal spoilage. Spoilage is caused by the presence of oxygen. To reduce the amount of damaging air and save a lot of feed that might otherwise be lost, a tractor runs back and forth across the pile during the silo filling.

During the 1940s, a trench silo was constructed on a farm in Crawford County, Pennsylvania. It was a pit type and was filled with long grass herbage cut in the field, which was raked, wilted, hauled to the silo without being chopped into short pieces, and blown into a box wagon. The grass was gathered with a buck rake and dumped into the pit, saving the use of a lot of excess machinery. During the winter, enough silage was sliced off with a hay knife to feed the cattle for one day. This took about three minutes. It was then loaded into a trailer and pulled into the barn for the cattle. Because of cattle habits, it was several days before most would accept this grass silage for feed; a few animals never did accept the addition of grass silage as a hay supplement for their meals.

If you should have enough acres in meadows not only to make enough hay to feed the year around but also to fill your silos with haylege or grass silage, proceed with this plan. Fill your silos with the above and then keep your cows in the barn at night and feed them. This grass silage will carry you through until you are ready to fill your silos with corn (silage) and hay through the winter months.

With the above plans, very little concentrate is needed. Your cattle will do just fine with a mixture of equal parts oats, wheat, and corn with a little soybean oil meal added (about one handful).

It is truly difficult to choose between grass and corn silage as feed for cattle. A milk tester who spends a lot of time with many farmers has the opportunity to see both feeding crops in action, and observations bear out the preference of corn over grass silage. Farmers prefer to feed their livestock grain at this time and place it on the silage, thus making the silage more palatable.

The grain mixture used with a good quality of roughage is usually much lower in protein and consists mainly of home-grown grains, ground locally. Possibly there may be a small amount of soybean oil meal added to balance the extra protein replacement the milk cow will lose during her lactation period, that period being the length of time a cow will milk during one year.

A basic grain mixture used by most farmers (per ton) for a grist would include 600 pounds of corn, 600 of oats, 600 of wheat, 200 of soybean oil meal, and 20 of salt. A good manager/farmer knows his cattle as individuals and will feed them as such; after all, their likes and dislikes are different. Raising any animal or herd of animals is rather like raising a family of children, the difference being that the animals mature much faster. When feeding grain to cattle, care should be used in the amount allotted to each animal; NEVER overindulge. Nonmilkers and low producers should be fed about four quarts of grain daily, while heavier producers should be fed no more than six to eight quarts regardless of the amount of milk produced. Baby the cattle and feed them individually as they are being milked. There should always be fresh hay in the manger while they are in the barn.

Never change the ingredients in the grist, but experiment with various grain mixtures and find a grist that they all seem to enjoy and stick with it. Here is one recommended feed for dairy cattle:

- 800 lbs. ground corn
- 800 lbs. ground oats
- 200 lbs. wheat bran
- 200 lbs. soybean oil meal
- 700 lbs. blackstrap molasses (approximately)
- 20 lbs. salt

These ingredients should be ground fairly coarse, as cattle prefer crushed rather than finely ground grain.

The dairy cow should have access to the best mixed hay you can place before her, but do not feed too much roughage at one time, as she will toss it around and scatter it so that it can become contaminated and unfit for consumption. Feed her in limited quantities but several times each day.

Here is an excellent program for taking care of your milking herd throughout the pasture season.

Always feed hay year-round regardless of how excellent your pasture is. Cattle will always eat dry roughage. This will help to lengthen the grazing season and also provide you with not only your seasonal increase in production but also a more contented herd. The better grazing for a much longer period of time than usual will hold your seasonal flow of milk straight through the summer months, providing you with more income from your milk.

To go one step further and show still more profit plus contented cows, keep your cattle in the barn at night. You will know where they are, and especially during thunderstorms, they will be dry and protected from the wrath of Dame Nature.

By keeping your cattle in the barn at night, you will benefit from all of the manure provided, which otherwise is misplaced in the pasture, where it is more of a contaminant than a benefit. By multiplying the bedding and manure mixture from each night's accumulation, you can top-dress a good many acres of your meadow and crop land, thereby enriching and building up the soil that otherwise would be lost. This is an excellent management practice, and results on the profit side are invaluable.

By following the above procedure, I found that I did not need to follow a rotational grazing practice. My cattle would graze until satisfied, then leave the pasture, returning to the barnyard where they would lie down and chew their cud until midafternoon, whence they would return to graze once more before chore time. Thus, there was much less contamination by their droppings in the pasture and more droppings near the barn where I could clean them up, which is like money in the bank.

Always remember that cattle will not do as well both physically and profitably if you change their daily routine abruptly. They become addicted to a daily routine and can be quickly upset if you try to change it. You will very quickly find out that you are not their boss but that they are actually running the show.

Chapter Eight

CATTLE HEALTH

When discussing the health of any living thing—animal or vegetable, human or livestock—you must begin at the source of all life, the soil.

Farming today is comparable to our industry. It is geared for high production at low cost, disregarding the quality of the product itself. Quality has become secondary to quantity and expediency. This is obvious in nearly everything we produce: have you ever looked at a modern car with its tinny construction and inferior materials and felt agreement with the old cliché "They don't make 'em like they used to"?

It's true. They don't make 'em like they used to and can't make 'em like they used to because of the need and desire to produce more and more of everything.

The same thing applies to the foods we eat today, although it is not as obvious to the untrained eye. You can't look at a piece of meat or a vegetable and tell that it's been mass produced, it has undergone a lot of treatments and processes that, if you could see them being performed and witness the long-range effects on your body, would doubtless make you a bit hesitant to eat the food set before you.

Most of the food we eat is stuffed with preservatives and chemicals. Just read the label on a foodstuff sometime and see how many natural ingredients you can find listed. Very few, I'll warrant. The long list of chemicals does not bother most of us, however; we shrug it off and say that they are a necessary evil. While I will agree that they are an evil, I cannot concede that the pollutants in the food we eat are all necessary.

Many harmful chemicals reach our food through the natural growth process without being directly added by the food industry. These are the chemicals that plants absorb when grown with inorganic fertilizers. It cannot be avoided; if plants are grown inorganically, there will always be extraneous chemicals present in them when they are havested. Some are perhaps harmless, but most are not. The plants are weakened (unhealthy).

Fertilizer may be a great boost to the plants to which it is first applied, making them grow faster and appear to be more productive. This is a superficial benefit, however, when damage to the soil, livestock, and —ultimately—people is taken into account. Fertilizer is nothing more than a short-cut to unhealthy farming and one that does more harm than real good.

Suppose a dairy farmer is grazing his herd on pastureland prepared inorganically. He coats the soil with chemicals that are absorbed by the plants, which, in turn, are absorbed by the cattle, which finally produce chemically lacerated milk. It stands to reason, doesn't it? Chemicals, once added to the soil, do not simply disappear. They end up in the milk and meat we feed to our children.

The chemicals present in most fertilizers directly

affect the health of livestock raised on inorganically prepared land. It makes them less productive and more susceptible to disease. Milk-producing animals raised by inorganic methods invariably contract acetonemia (ketosis) after freshening. They also are vulnerable to milk fever. Both of these illnesses result from an imbalance of trace elements found only in healthy soils. Once an animal contracts one or both of these common diseases, the services of a veterinarian are required to bring the animal back to health. If you have checked on the cost of hiring a vet lately, you know that this is a crushing expense to a farmer. Yet it is not an expense that the farmer needs to suffer; it is brought about by incorrect farming methods. It is the price that inorganic farmers pay for their ignorance.

The worst illness that afflicts cattle is mastitis. This disease is a killer and will wipe out calves, heifers, and mature cattle at random. It can destroy a farmer's herd if unchecked.

For years, these diseases have been discussed by the know-it-alls of inorganic farming, the college university professors with degrees in animal husbandry. They give various explanations of ways to prevent the disease, none of which work. For all their education and their sophisticated research, they never seem to think that inorganic fertilizers are really the culprit. They can't see the forest for the trees and overlook the most obvious cause of animal illness: chemical pollutants.

Not only college and university professors but veterinarians as well do not understand the cause for mastitis, so they are still in the dark as to how to treat it. If the cause were explained to them, they still wouldn't believe it. This pertains to our medical doctors and scientists as well. They are spending multimillions of dollars for cancer, diabetes, emphysema, and heart disease cures and still are stabbing in the dark for answers. Patients being attended by doctors today are nothing but a bunch of guinea pigs. The doctor will tell them to try this prescription for a week and, if it doesn't help them any, to come back next week and try something else!

A healthy animal will not contract mastitis, acetonemia, or milk fever. These are diseases of unhealthy cattle. I have never seen an article on these diseases, "illnesses," that even came close to knowing the cause, effect, or cure. And I subscribe to several farm periodicals per month.

I do know that there is a lot of not only money but also power in the drug companies, and this applies to both humans and animals alike. As long as the public can be kept in ignorance and bled white of their money, why should a cure for these diseases be formulated?

A cow that has a chronic case of mastitis, one that affects one or two quarters of the animal's udder, is really sick because her system is full of the bacteria that causes the congestion in her udder. The udder is feverish, swollen, hard, tender to the touch, etc. If the feed she is eating is not stopped or drastically changed, she will end up at the slaughterhouse in spite of all the veterinarian can do.

Mastitis is not contagious to a healthy animal, so unless you are farming inorganically, you will never have to worry about mastitis or any of the other above-mentioned diseases.

In most of the articles I have read on mastitis and its prevention, the causes all seem to be about the same; drafty barns, cold, damp floors, too much vacuum in the milk line, teat cups left on the animal too long. None of these causes seem to bother a healthy animal!

The preventive measures for mastitis in these articles seem to be about the same over the years. Use teat dip before and after milking, permit a few squirts in a strip cup before placing milker on animal to be milked; when animal has completed a lactation, give each quarter a treatment with antibiotics, etc. These are more preventive measures than cures; usually, a veterinarian is called in for consultation, and he gives her antibiotics, etc. The chances are that in a good many of the chronic cases of mastitis, the animal will usually lose that quarter for any further production during her remaining production years.

You would be surprised to know the number of cattle that have just freshened and in less than a month of their lactation were slaughtered for beef. This beef is not healthy but is being killed and processed for human consumption every day. To go a step further, you might be surprised to know the number of calves, heifers, and cows that were hauled to the woods and buried that never arrived at the slaughterhouse.

We live today in a drug-based culture, and a person is a fool who cannot realize it. Just look at the ads on television: you can get pills to cure everything from insomnia to acne. Most of them don't work, but people still buy them by the cartload and cram them into their systems with the belief that these drugs are an instant cure for anything that might be wrong with them. Technology has run amuck, and medicine has become one of the most dishonest aspects of our society.

The same thing, the same sickness, reaches into the farming industry. You can buy additives, medicines, pills, and all sorts of ridiculous contrivances that promise to keep your animals healthy. People, for some reason, trust these drugs and use them. We use chemicals to destroy our soil and then think we can undo the damage simply by adding a few more chemicals to the animals and crops we raise.

Farmers today have ketosis in their herds so bad that you can smell it before entering the barn. It smells like a sickroom, like medicine. The milk from these herds has the same medicinal taste and aroma.

This results directly from the use of inorganic fertilizers. Sickly soil, sickly crops, sickly livestock, sickly people, they are all a part of the inorganic method.

Chapter Nine

CALF HUSBANDRY

Every farm animal is important to a dedicated farmer. To a dairy farmer, calves are especially important because they comprise his future milking herd; these calves will someday replace cattle the farmer loses due to low production, old age, sterility, or some other organic cause.

Most cattle diseases are the result of inferior feeding programs that utilize the poor crops harvested from unhealthy soil; once again, the culprit here is artificial fertilizer. Constant use of this fertilizer could eventually destroy the entire earth. During World War II and years afterward, all commercial pig and chicken feeds on the market contained drugs, such as antibodies, etc. This may still be true in much of our country today; we will see how such chemical processes can greatly harm our farm wildlife.

Each farmer has his own unique way of raising calves. As a supervisor of the Dairy Herd Improvement Association some years ago, I had the opportunity to observe and evaluate many methods of calf husbandry. This was an extremely educational, interesting, and beneficial experience in that it helped me realize the best possible way to raise calves with the best results. I now use what I consider to be the easiest, most efficient and closest-to-natural method. I have seen superb results and have decided beyond the shadow of a doubt that it is the best way to raise calves.

A farmer, when raising his own calves, must first turn his attention to the dam (mother-to-be). The condition of the calf's health depends on that of the dam, so she is the immediate concern. If the farmer farms organically, there should be no problems; if, on the other hand, he follows inorganic processes, then both the cow and the calf may encounter serious health problems.

On an inorganic farm, the dam's general health will most likely be below normal, with a high probability that a majority of such animals will have other serious illnesses every year. Inorganic farmers can probably expect the usual inorganic-related diseases, such as mastitis, milk fever, and acetonemia; however, there also may be several birth complications that can be related to poorly planned farm practices. A slow or prolonged birth can result from poor tissue due to lack of trace minerals not available on inorganically operated farms. Often, the afterbirth tissues will not disengage, thus making dry births quite common. Both deformed and stillborn calves may be born too frequently, also; synthetic fertilizers deprive the plants of natural organic trace minerals and, likewise, deprive the cow of some crucial nutritional needs. Feeding high-producing animals with hormones and/or drugs can easily destroy the fetus.

After the calf is born, assuming there have been no major complications, it is better to remove it from its mother soon; the longer the calf is under its mother's care, the longer it will take the cow to settle down and increase her production. It is also more difficult to train the calf to nurse the way you want it to, so it is better to have it on its own as soon as possible.

The first and most important need in weaning a calf is SANITATION; make a special effort to be sure that all things in contact with the calf are absolutely sanitary. Be certain the pen in which you place the calf is perfectly clean, contains several inches of fresh bedding, and shows no signs of draft or dampness. Replace the bedding daily with equal amounts of new bedding, being certain to remove all calf droppings from the pen each day—another simple sanitation matter.

Calf nursing and feeding can be a delicate subject. Do not force the little animal to nurse and do *not* overfeed it! It will eat when hunger necessitates, a perfectly natural means for existence. Depending on the size of the calf, the liquid given to her at each feeding should not make the paunch (stomach) extend beyond a quarter moon in size and appearance. A forty-five to fifty-five-pound calf should be fed approximately one quart of milk twice each day, and this amount should increase proportionately according again to the calf's size. There is a special diluted milk formula that should be maintained for the first ten to fourteen days. After being sure all utensils are clean, dilute the mother's milk with equal amounts of water (50–50 percent) if the milk is from a Jersey or Guernsey cow; this is because these cows have a higher percentage of butterfat in their milk. If the dam was a Holstein, the calf's formula should contain 75 percent milk and 25 percent water; after two weeks, the water should be increased to 50 percent of the total liquid. There also should be an automatic water bowl filled with fresh water at all times after the first two weeks.

Avoid using a milk replacer; more calves are fed on these than on raw milk. The calf should be fed the aforementioned formula, utilizing milk from its mother and water. Calves can be best fed with a nipple pail, a ten-quart pail with a four-inch rubber nipple attached near the bottom. The amount of milk the calf gets is regulated, and also provides the necessary simulation of natural nursing.

As the calf grows older, increase its amount of water in the milk formula until, at about four months of age, it is drinking only water daily from the nipple pail and remaining on water for about ten to fourteen days. Then cease feeding with nipple pail. By this time, the animal should be using only the water bowl.

You can feed the same concentrate formula to your calves as you do your cattle. With the addition of calf manna, which is extremely palatable, they will come along nicely.

A good calf grain may be placed in a grain box in the calf pen right away, and the calf will usually find it within a few days. Provide only enough for the calf to clean up during the day. Fresh, sweet legume hay should be placed in a manger available to the calf after it reaches the age of four to six weeks. There is a wonderful product called Calf Manna that is a superior pelleted concentrate to feed the calves as a supplement to their daily grain. It is a perfect concentrate and contains all of the necessary trace minerals for a growing calf's diet. Until the calf is weaned, feed this supplement in with the regular grain.

Because calves are individuals, they may occasionally go off their feed. If this happens, beat two eggs well and mix with the calf's milk, which should be decreased to about half the normal ration. This will usually, after a couple of feedings, put the calf back into its routine; then normal feed can be increased to the standard level.

Remember that all young animals, even human babies, have a natural instinct to chew on just about anything chewable. In the case of calves, this involves a wood-chewing habit that is no need for concern; even adult cattle lick or chew on wood, especially if it is a soft wood. Veterinarians, professional agriculturalists, and salesmen from trace mineral companies write and preach that this trait exhibits a need for trace minerals in the calf's diet, selling the farmers a lot of worthless and unnecessary mineral supplements. The calf will prove that it does not need them.

In inorganic practice, the illness that causes the highest mortality rate in calves, especially in those still receiving milk in their diets, is calf scours. This is caused primarily by overfeeding, feeding too much rich milk and letting the calf eat when its digestive system is upset. Remember that the calf raised under inorganic procedures will more likely require antibiotics in its feed to prevent scours, while the organically raised calf will not; the former calf will be more susceptible to illness and disease than a calf grown on an organic farm. Unless you have practiced both farming methods, the difference seems minimal. But seeing the tremendous improvement in raising animals on an organic farm after doing so on an inorganic farm is almost like being able to see again after being blind!

Try this method; your cattle will thank you for it.

Chapter Ten

SOME PROBLEMS OF THE POORLY EDUCATED FARMER

In chemical farming, the end result is always disease, first in the land, then in the plants and animals, and finally in man. Wherever this type of commercial farming is practiced, people are sick! In this case, the frequently used phrase "You are what you eat" is appropriate, for eating foods full of chemicals is bound to present later problems. The higher the organic fertility of the soil, the better the health of animals and man will be.

Today's farm and gardening publications, one of the largest influences on the farmer, seldom circulate articles concerning the healthful aspects of organic farming practices. This issue, in fact, seems to be a minor one for agricultural and science educators. Articles written by supposedly brilliant men—professional farmers and laboratory techicians who provide examples for the public—seem to cater to the manufacture of commercial fertilizers and drugs. Popular agriculture periodicals deal primarily with antibiotics, herbicides, insecticide sprays, etc. Soon it becomes obvious that this is a vicious cycle, for the more we use these products, the more we will need them later as we continue to poison the environment.

These learned men are actually brainwashing the vulnerable farmer, who follows this chemical trend rather than pursuing his own instincts or background: to farm organically. In seeking advice on the general care of crops or fruit trees, farmers are told to rely on expensive, poisonous fertilizers and herbicidal sprays. They do not stop to consider that these elements, meant to repair, are absorbed in the same way as crop-nourishing materials; it is easy to see how all forms of life coming in contact with these inorganic substances will suffer.

Due to rapidly changing and so-called improved methods of farming, the farmer often feels he cannot keep up with the times. A powerful trend has been established and farmers, like everyone else, feel they must follow it to succeed; it seems that all these modern methods are "the thing to do." Agriculture has become an advanced science and has moved into the educational field as well, so that when a farmer needs advice, he turns to universities or cooperatives. Today's farmer does not really operate his own business anymore but depends on behind-the-scene agriculturalists. It is these "armchair farmers" that give advice to today's farmer; what to plant in which acres, how much and what kind of fertilizers to use, and the proper use of these chemicals. The dependence of farmers on these "experts" will be the downfall of American life as we know it.

One weak area resulting from relying on advanced farmers involves the seeds the farmer plants each growing season. In the last thirty-five to forty years, hybrid seeds have been used extensively in both farming and gardening. By definition, a hybrid is "the end result of anything of mixed origin." New varieties of seeds and plants through hybridization have been a marvelous discovery—to a certain point. However, we have now become dependent on this mode of seed culture. This means that a farmer cannot select seed from each autumn harvest to plant the following spring; it will either revert to a wild state or to a crop that would be a great variation of the previous seed. When this happens, the farmer must purchase his seed from retail sources, an expensive annual proposition. Also, in the event of a natural catastrophe, such as a drought or insect plague, the harvest could be ruined, resulting in a severe shortage of seed for the next spring planting. This, in turn, could cause widespread famine and worldwide starvation.

Several times during the past century, this country has secured new varieties of seed (cereal grain) from Europe to compensate for our own low yield and diseased crops. European grains had always been raised under organic farming conditions; American crops have been almost completely influenced by inorganic practices—commercial fertilizers and lime. As we have discussed, these chemical substances act as a sort of cancerous growth, destroying or drastically reducing natural antibodies that protect the seeds or plant. In the human body, if the blood is diseased, the body will soon succumb to a fatal illness, and if the source of the sickness is not treated, there is no hope for a cure. In the same way, many organ transplants are rejected by the body because they are alien substances. Why should we not expect the same treatment with chemicals in natural soil?

The new European grain varieties were distributed

to farmers and planted in their *sick* soils. For several generations, they remained superior to American grains, but after continued exposure to lime and commercial fertilizer, they deteriorated to the same low quality as our previous seeds. We would see tremenddous physical damage by delving into the deeper phases of biological effects not only on plants but also on the soil, water table, and animal life, all of which have been in contact with artificial fertilizers.

In short, these inorganic materials deplete the soil of its biological life. The plants' nutritional necessities become leached out, soon making the soil sterile and barren if not fed copious amounts of organic substances.

Therefore, it is crucial that farmers retain an open mind and not readily accept everything that the experts in agribusiness advise. How do you know what to do, who to believe? The simplest answer to that is to begin to farm organically. New and improved machinery and methods can still be used in organic farming practices, but the results will be incredibly different! Why spend money and time on products with a failure risk greater than ever before? It is the long-term results that people often overlook in examining organic methods; they consider the tremendous increase in quantity, when most important should be QUALITY.

So it is crucial to THINK when you embark on a farming experience. Be sure you feel you are doing the right thing, that you would gladly eat what you have grown or could sell it as a healthy food to someone else. Think when you read about that medicine for your dairy cattle; will it somehow affect the milk or harm the cow? What about all those sprays and plane-spread insecticides? Surely, they are doing more harm than good. Read about the widespread disease hitting plants and farm animals. A careful study will reveal that inorganic farming is not the answer; dangerous chemicals are not a good cure for what ails us.

Soil farmed organically is full of both "death and life." Death and decay of organic matter contributes to life in the form of molds, bacteria, and earthworms. This, in turn, creates healthy organic food for the farmer's crops. It is not too late to reeducate ourselves in the area of agriculture, but it must be done NOW, for the world's welfare depends on it.

I read an article recently where a well-known agriculturist made the remark that we cannot afford to turn the clock back and return to former farming practices. That we could only afford to forge ahead. Future progress was the only answer.

If you will follow the procedures outlined in this book, you will very easily outproduce any inorganic farming methods tried in the past, now, or in the future.

Chapter Eleven

WHAT TO LOOK FOR WHEN PURCHASING A FARM

Suppose a person wanted to buy a farm and go into the farming business. What things should he be thinking of when he makes his purchase?

There are several major considerations to be made. One, obviously, is the type and condition of the soil. Another involves the buildings on the property: are they in good condition, and are they suited for the type of farming the individual wants to undertake? As for machinery, which is most often sold with the farm, is it what's needed to do the job? And how large a farm should it be? All these questions should be thoughtfully considered.

The size of the farming enterprise is dependent on whether you are a dairyman or a beef cattle manager. The type of farming that is prevalent in the neighborhood where you are considering your operation is important.

For a beef cattle operation, a larger acreage is permissible because grazing is a year-round operation regardless of weather conditions.

For a dairy operation, the farmer does most of the

harvesting, and the feed is brought to the cattle; therefore, the size of the business is regulated by the number of milking cattle in the herd.

Where the winter months are harsh, and there could be an accumulation of snow for an extended period of time, beef cattle must be fed most of the roughage that they consume during that time.

Dairy cattle are fed the largest share of their roughage in or near the milking barn. Milk cows cannot stay out of doors in below-freezing temperatures for any length of time, as their teats and udder will freeze, gangrene will set in, and they will be lost to the milking herd.

The average young man wanting to start in the dairy business today would require capital in the neighborhood of fifty-thousand dollars. Where can a young married man find that kind of money? It is extremely hard to start an operation of such magnitude. You should be able to ship your milk to a cooperative immediately so as to have an income. You can presently see if this were possible it would mean that this farm should be classified such as A, B, C, or D. By that, I mean, it should have a barn full of milking cows, the milk house should be inspected and okayed, the meadows should be seeded to mixed grasses and legumes and ready for harvest, there should be plenty of good water in the pastures, etc.

There are several ways in which you can achieve this ambition: one is through inheritance; another way is through partnership. Still another way is by good fortune: someone who has money and will invest in you as a person, who has faith enough in you to realize your dedication and your need.

Climatic conditions also could be a question as to

the type of farming enterprise you are best suited for. Moisture is the outstanding asset. In the Midwest, where moisture is the determining factor between a successful crop or a failure, your investment might be risky. Here in the East, where we have plenty of moisture, your venture could be successful.

If you want to purchase a farm and are not particular as to the soil classification, you may have to be concerned as to the type of farm business you will be undertaking.

By classifying the soils as (a) clay, (b) loam, (c) gravel, and (d) sandy loam, it would be a better choice to choose a clay-type soil for grass farming, dairying, and beef cattle production. A loam soil would be the best all-around soil, especially suited for hay, dairy cattle, beef, grain and root crops, and market gardening. A gravel soil would be excellent for the production of grain, dairying, and beef cattle production, and a sandy loam soil would be a good choice for root crops, market gardening, and orchard with a provision for irrigation.

To renovate a supposedly worn-out soil properly, there are two preferable ways:

One, use the proper tillage tools, such as a subsoiler and a field cultivator. Then top-dress liberally with a coat of green manure, followed with a seeding of a perennial grass as a follow-up green manure crop. In the spring, disc this fall seeding in with the soil, then use this field with your own discretion.

Two, use the same tillage tools as above; then seed it to a green manure crop, such as perennial rye grass if you do not have the manure for top dressing; then, the following spring, disc the crop of rye grass in with the soil and if it has mellowed the soil sufficiently, and it looks enriched enough, either seed it to a good hay mixture or to a grain crop, such as oats, where you can disk the straw into the soil before using it to your own discretion.

After the soil in your fields has regained its former natural and rich texture, you can keep it productive beyond all expectations by just top-dressing it with a medium coat of manure biannually.

Some of the greatest cows I have ever come in contact with were grade cattle. Yet to obtain a herd of pure-bred animals is a great achievement. The term "pure bred" implies generations of a freedom from mixture or cross.

I would neither specify breed of cattle nor whether they were grade cattle or pure breds. It is their production and health that counts.

The selection of tractors and related tillage equipment is very important, expecially if you are short of help. Never purchase a tractor and equipment too large for your business. This will run your expense of operating this equipment too high for your size of business.

I prefer a three-point hitch tractor with as much three-point-hitch adaptable machinery as possible. They are easier to transport and are less expensive, do nicer work, and are easier to maneuver.

If you are in a farming area, you possibly would not have to invest in all of the machinery you might need for your diversified or general farming because what you don't have in the complete line of farm machinery your neighbors will possibly have, and you can trade work with them when you need the use of a piece of their equipment.

To Paul Pfund—and with all sincerity I say:

I cannot remember when, where, or how I found this magnificent statement of faith, principles, opinions, and truth so descriptive and beautifully written and for such a great movement as man is now undertaking. I give praise to this man with the deepest humility for putting into words these God-inspired beliefs. I sincerely thank Paul Pfund from the bottom of my heart for—"My Creed of the Soil":

> I believe in the soil as God's richest and most bounteous gift to man. It is the sustainer of his physical being which is the embodiment of His spirit. It abounds in life in its most productive and complex form, yielding unto all who wisely culture it, a benevolence and benediction which can be ascribed only and alone to the all-wise Creator of the Universe.
>
> I sense in the soil an inherent quality so vital and powerful and yet so delicate and sensitive as to merit my best thought. For, as in the measure it is truly preserved, so shall man be sustained. From it springs all forms of life and by it are all nourished. None is exempt from dependence upon it.
>
> Gratitude and appreciation well up within me for such an abundant provision. I feel a deep sense of responsibility to my Maker to do my utmost in exercising proper diligence as keeper of a sacred trust and the discharge of a faithful stewardship.
>
> I, THEREFORE, RESOLVE TO: Commit myself wholly to the principle that the Supreme Being has made full provision for a complete, wholesome sustenance of life within the soil.
>
> Never knowingly desecrate its matchless precincts by any action or with any substance in any manner or means incompatible with its immutable

biological nature. Avoid most precisely, as a matter of practice, the use of unnatural lifeless stimulants intended to enrich the soil, for only that which has life or its qualifications can impart life.

Avoid the use of any baneful insecticides and fungicides, using if necessary, only non-poisonous repellents.

Consistently study to show myself a trustworthy keeper of the earth, wisely contemplating most carefully all my actions in relation thereto.

Consistently practice essential conservation by composting all suitable residues for soil rejuvenation, and enrichment, thus preserving the soil which I construe to be the true measure of cultural and abundant living.

EPILOGUE

This is not a very large book; for its purpose, I deemed it unnecessary to camouflage the material written herein among several hundred pages of nonrelated words. It would make it rather difficult to pinpoint the subject material contained here, which is the very purpose proposed in the first place.

The subject material that I have gathered for this book is from a lifetime of knowledge and was planned, experimented on, and sometimes carried out through trial and error. It is knowledge gained through experience.

I could have interspersed several chapters on other related subjects but decided in the negative.

I have subject material ready for a book on organic gardening but thought I would wait and see how this book was received. In other words, first things first.

There must be a change of heart in the stewards of the soil, or starvation through malnutrition and disease through malpractice in agribusiness will become a reality the world over. Therefore, I do believe the time is ripe for the message and methods contained in these chapters. If this book is not a success, *God Help Us!!!*